Copyright © 2023 by Richard Wain

All rights reserved. No part of this book may be reproduced or used in any manner without written permission of the copyright owner except for the use of quotations in a book review.

First paperback edition Nov 2023

Book design by Richard Wain

ISBN 978-1-7394516-1-5 (hardback)
ISBN 978-1-7394516-0-8 (paperback)
ISBN 978-1-7394516-2-2 (ebook)

Published by
:(LITTLE DRAMAS PRESS :)

www.lifeslittledramas.com

To Asher and Jowan

For the future they deserve.

Foreword

We're on the edge of change: The Great Transformation, Great Adaptation, Great Unravelling, or just a Great Deal of Uncertainty With Which We Are Currently Not Coping. If you're reading this, you have some idea of the depth of the crisis. You may lose sleep to climate anxiety, or watch your children as they realise their future is not the one we imagined for them when we were their age. Whatever your emotional steps into this, what you almost certainly don't have is a road map for navigating this level of personal and cultural insecurity.

What we all desperately need is a sense that we are not alone, that we have fellow travellers who share the dark nights and moments of despair, but who can also unearth from the dark, nuggets of raw compassion, of insight, of self-reflection and, dare we say it, hope, even if it's the Vaclev Havel kind of hope that is not the conviction that things will turn out well (because at this stage, that would be delusional), but a belief that things make sense, however they turn out.

These poems offer this kind of sense-making, from the Introduction where you reach the cliff's edge and realise you are not alone, through the rhythmic promise of an unravelling and reweaving in The Future is Already Here, to the wonder of engagement in Atoms and the joy of discovery through the telescope in Beyond to the implicit community in We Share a Dream... each of these pieces offers new insight, new possibility and the sense of co-creation that is at the heart of all the best art.

I share with Richard the belief that there is still time for a decent future to emerge from the collapse of the current system: that we can leave to our grandchildren's grandchildren a flourishing world, where people and planet thrive, where humanity steps into the niche that the web of life offers us and where each of us brings the best of ourselves to the collective table, confident that we are a thread in the overall weave and this weave is beautiful.

If we're to get to this, we need encouragement to let go of the old paradigm and step with joy, courage and conviction into the new. These pages will help pave the route. I commend them to your hearts and your souls.

Manda Scott, *author of the Boudica: Dreaming series and host of the Accidental Gods podcast*

My world is changing.
Can I change too?

CONTENTS

Introduction - 8

The Future is Already Here - 14

Lose Myself - 18

Atoms - 22

She Learned How to Dance - 26

Beyond - 31

Row Across the Sea - 38

The Letting Go - 42

Beyond the Brink is the Beginning - 48

A Song from the Edge of the Universe - 52

We Share a Dream - 56

In the Eyes of Another - 62

What Next? - 64

Travellers Beyond the Brink - 65

Introduction

Feel the breeze now,
steady and cool against your skin,
your senses raised
to greet the chill.

The early autumn leaves
shift uneasily
as if in memory
of some unspoken doubt.

Dew soaked ground.
A whitewashed sun,
reluctant, somehow
to embrace the jagged horizon.

Each step reveals
the broadening sky,
grizzled trees give way to shale
as the old world falls away.

The chasm before you dives
sheer into mist. Lonely islands,
like driftwood on a shifting tide,
dance with the waves and wisps of shadow.

There is no way down,
no path to lead you back.
You are at the brink now.
You have nowhere else to turn.

Eyes drawn to the high hills,
the distant horizon,
impossible thoughts,
curving light and shadow.

Your heart beats
faster now.
It beats in anticipation
of what lies beyond.

You can feel the danger
there in that untold wild,
the hunting eyes,
the webs of cruel uncertainty.

But you feel something else too,
something that calls you,
something that knows why you are here.
Something brave. Something new.

And as you reach
the cliff edge
you realise
that you are not alone.

There are other people
stepping out
from the tree line.
So many people,

moving as one,
so different, yet the same,
shielding their eyes
as they step into the light.

They reach the edge,
look over and exchange
concerned murmurings, nervous
glances.

More and more step forward
until the mile long curve
of the escarpment edge is hidden
entirely by the gathered crowd.

All fall silent.
As one
they look out
to the far away mountains.

A sense of knowing
and not knowing
passes between them.
The knowledge of an ending.

But more than that,
a dawning awareness
that beyond that ending,
beyond the brink, is the beginning.

The beginning of a new world.
The ending of the old.
Standing there at the edge,
everyone can see it.

If I step forward,
what kind of future
will I find?

The Future is Already Here

The future you seek is already here
It's locked to the beat of your heart
When the world that we weave is unwoven
every thread that is left plays its part

In weaving a new web of meaning
In planting the seeds of belief
A vision that follows the dreaming
The root and the branch and the leaf

All patterns are interconnected
Each fractal a sum of the whole
Our world in its chaos and glory
A body at one with its soul

Its true source of power is compassion
The calling so gentle and clear
A whisper of wisdom and wonder
For the future that's already here

Do I pay attention to the beauty in this world?

Lose Myself

Right here, in this moment
bells chime for newlyweds
Wood pigeon calls like
an echo from an older church

The road narrows, claimed
by roots and needle dust
Deer print earth, leading
down a fern lined aisle

To the altar stone, hewn
by fallen rain and time
Dippers in their dinner jackets
dance with nymphs and shadows

Leaf music thrills the breeze
between these towering pillars
Embracing earth and sky
Their love unbowed by years

Whose ballad, lost in time
still lifts each beating heart
Boughs tremble as we dance
I lose myself in your arms

Am I alone
or am I part of
something bigger?

Atoms

Each of us wakes in the morning
Wakes up in a body alone
A flicker of consciousness burning
Encased in our flesh and our bones

Each one a collection of organs
In turn a collection of cells
Each built from a billion atoms
With a billion stories to tell

Some from the stars up above us
Some built of sand and of stone
Inside every one of our bodies
A million dinosaurs roam

We're each made of Julius Caesar
We're each made of Emperor Qin
A little of William Shakespeare
A little of Anne Boleyn

And ten million years in the future
Each atom will surely still be
The tiniest part of the body
That long ago used to be me

New life will evolve from our ashes
Composed of the memories lost
Some made out of Brittany Spears
Some made out of David Frost

Some made out of Marcus Rashford
And some out of Meghan Markle
And then your time will come again
Another chance to sparkle

As the atoms that were you
Are now in everything alive
You are everywhere at once
Every piece of you survives

Not alone within your body
But out in the world and free
There is no way of knowing
All the wonders that you'll be

Future worlds that you'll discover
Out of time and out of mind
Freed of consciousness and body
Heaven knows what you will find

When you fly beyond the mountains
Beyond touch and taste and sight
You will transcend this mortal life
Before you say goodnight

What will my Grandchildren's lives be like?

She learned how to dance

She was born yesterday
And grew through the turning
The tide, ever changing
A time beyond yearning

She watched as the men
Who claimed the old ways
Let go and stepped into
These halcyon days

She learned how to dance
To the most ancient song
As the farmland grew rich
And the wild woods strong

And she learned how to lead
With her sisters and brothers
Who'd meet eye to eye
To break bread with each other

And to use what they knew
To each unlock a door
To a new way of being
New paths to explore

New realms of potential
New hope against fear
Possibilities arc
Bent into a sphere

This world she inhabits
A blue pearl on black
Once close to its limits
Has found its way back

And now she is old
She sits in her chair
As her grandchildren play
She breathes the clear air

And slowly her face
Breaks into a smile
She closes her eyes
Rests there for a while

She dreams that her story
Rolls back through the years
That her grandparents knew
They would not fall to fear

That she gave all she could
And they gave her that chance
That she shared in their song
And she learned how to dance

Can I look beyond myself?

Beyond

A box sits on a table
I'm excited to get in
I take a pair of scissors
Cut the tape and then begin

To pull it open at one end
To take out all that lies within
Small boxes, metal legs
A bracket on which it can spin

And then I lift the telescope
And carefully I place it down
Take out the instructions
And inspect them with a frown

They're in Chinese... but that's ok
Because they come with little pictures
And the diagrams explain
All of the fittings and the fixtures

So step one is where I start
And the pieces come together
And I step through every step
Until I've finished the endeavour

Until standing on the floor
In all it's three-legged splendour
A calling from another world
To which I must surrender

Beyond each breath another breath
Beyond each cloud the sky
Beyond each far horizon
Beyond all thought I fly
Beyond each star another star
And each encircling world
Beyond each spiral galaxy
The universe unfurled

If only I could focus
This infuriating thing
I'm not sure it's set up right
I can't move the focus ring

And there's a little movement
In the viewfinder so I
Can't really keep it steady
As I look into the sky

I feel my irritation
Gradually rise
This is supposed to be
A life affirming enterprise

But now the cloud covers the moon
And I contemplate defeat
It's getting cold outside
I make a glum retreat

Back down the garden steps
And in through the back door
No shining distant secrets
No cosmos to explore

Beyond each breath another breath
Beyond each cloud the sky
Beyond each far horizon
Beyond all thought I fly
Beyond each star another star
And each encircling world
Beyond each spiral galaxy
The universe unfurled

This time the sky is open
No weather of this Earth
Will block my views of heaven
Tonight will be the birth

Of an amateur astronomer
Of promise and renown
I scan the wide and inky sky
Admire it's glittering crown

I try to work out first of all
Where all the planets are
I think I can see Mars
I'm not sure where the others are

I try looking through the scope
But all I can see is black
I keep trying and eventually
I start to get the knack

I find a star and watch it drift
Across my line of sight
It's the same as from the ground
A bit like a... satellite

I swallow disappointment
And try an app that I've downloaded
It shows planets, stars and nebulas
And stars that have exploded

And space stations and galaxies
Orbits, constellations
Digitally recreated
From scientific observations

And as I move my phone around
I see a planet on the screen
I take a breath because I know
That I must see what I have seen

Not through an app but with my eyes
Not in a book or on TV
My heart is wondering the skies
I want to see

I have to see

And so I try.
My telescope
surveys the sky.

I find a dot
Of purest white,
like any other,
but as I magnify
discover...

So impossible
So real
So unexpected
And I feel
My breath held
My heart begins
To race
As I look out
Into deepest space
As I focus
On a distant world

A perfect sphere
Of yellow pearl
Around it wrapped
A disk of light
It's shadow cast
against the night.

In Saturn's rings I lose myself
And melt away all fears
There is nothing anymore
And nothing wipes the tear

That trickles down my cheek
As the planet drifts from view
I have opened up the box
I have seen the world anew

Beyond each breath another breath
Beyond each cloud the sky
Beyond each far horizon
Beyond all thought I fly
Beyond each star another star
And each encircling world
Beyond each spiral galaxy
The universe unfurled

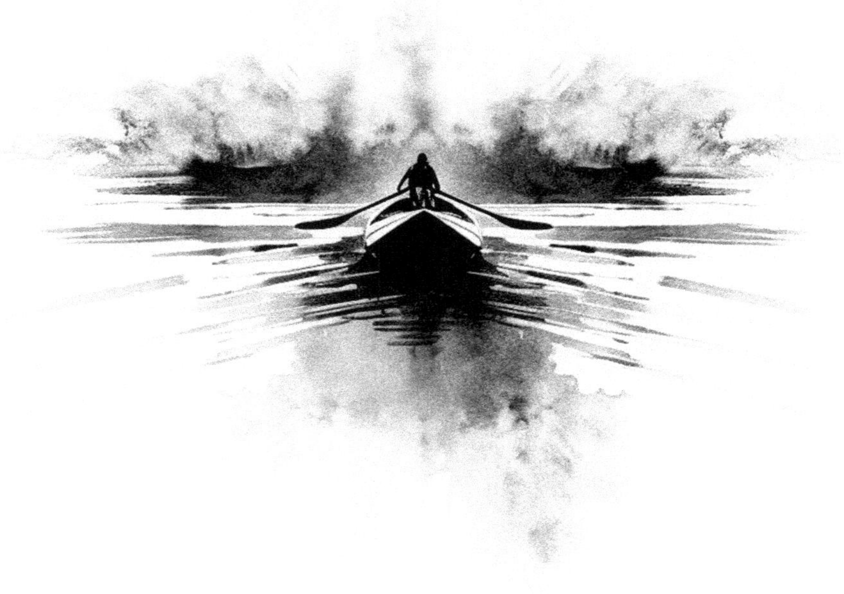

Do I have the courage
to take a new path?

Row across the sea

I row across this raging sea
From a land now far behind
When I set out in search of me
I did not know what I would find

I did not like the job I had
It was the paper pushing kind
It never moved beyond my wallet
To my heart or to my mind

Spreadsheet slaves in tiny boxes
Fingers drumming out a beat
To the tune of swift retirement
Or resentment or defeat

A beat so fast, the boxes filled
The pay cheque lands, I pay the bills
I watch TV till I turn grey
I don't want life to end this way

And so I row across the sea
To find a land where I can be
The person I was born to love
Whose dreams may fall from up above

Dreams like raindrops on my brow
A beat that lifts my heart and now

I hear the song that leads me on
My time is mine my shackles gone

I'm free to care, I'm free to grow
I'm free to practise what I know

Rewarded for all that I bring
No longer tied to all these things
I neither need nor want to be
A distant land where I am me

And so I row across the sea
Although I do not know the way
And I can feel deep in my heart
That I will find that land someday

I did not know what I would find
When I set out in search of me
But I will row and I will row
Until the rowing sets me free

I will row and I will row
Until I see that distant land
Pull my boat up on the shore
And set out across the sand

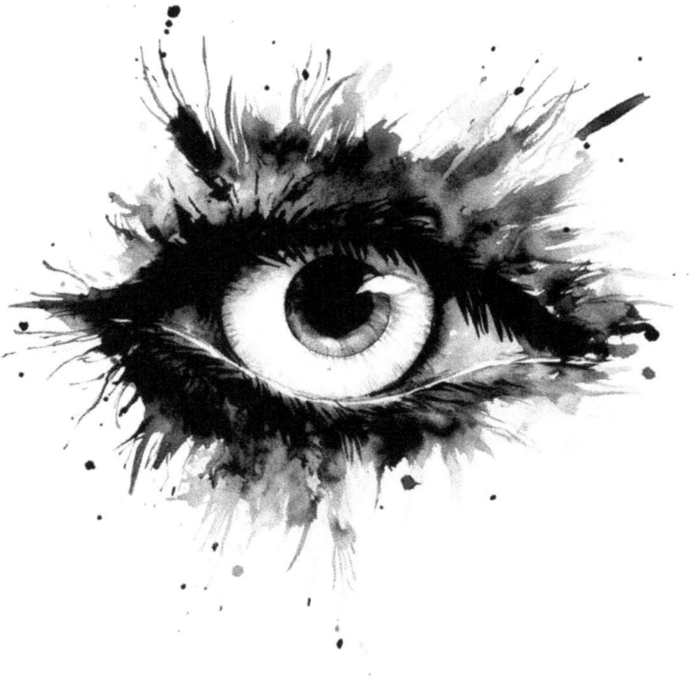

Is the wild still
inside me
somewhere?

The Letting Go

The letting go...

The letting go...

The letting go is not so easy
Though my purpose may proceed me
There is courage still to find

These twigs that bare me
Born from branches
Flesh unfolding from the soil
And I am here, clinging lightly
One swift breeze and I will fall

Twist and turn upon the current
Ride the wind to who knows where
All the earth laid out in greeting
Spinning gently through the air

I come to rest... a moment's peace...
The consequence of this release
This letting go, this setting free
A new perspective comes to me

As high above I see the branches
Charcoal slashed on paper clouds
These narrow paths that lead us skywards
All begin on solid ground

I sink my roots into the richness
Calling out to those below
Calling out to the invisible
We all must meet to grow

The Tardigrades and Nematodes
The Earthworms and bacteria
I call out to the fungi
To their networks of mycelia

Alone I starve, alone I fail
Alone this journey ends
There is no future for me
Without these trusted friends

And so I wait, and so I call
They do not come
And coldness falls

The waking up...

The waking up...

The waking up is not so easy
But I hear someone near me
I can feel I'm not alone

Slowly food and warmth revive me
There is something new inside me
Fresh green shoots of purpose bind me
to this soil I now call home

They heard my call and so they came
And now the flow of life is set
Time is nothing but attention
Free from pain and from regret

Time to live and to connect
Time to grab my chance and be
I can see into my future
I can see the towering tree

But first patience…
But first time…
Pay attention to the climb
There is no leap from here to there
That does not pass through everywhere

The slowing down…

The slowing down…

The slowing down is not so easy
With my friends now here beside me
There is wisdom still to find

As the years start to wonder
I reach deep into the earth
Silent hours left to ponder
This great cycle of rebirth

A new strength rises within me
As my branches now unfold
I recognise my own beginning
A thousand fingertips take hold

Of life's abundance bought to being
Future forests yet to rise
Yet to set root in the soil
Yet to reach towards the sky

Fed by webs of interbeing
By the seasons ebb and flow
By our mother, by our sun
And by the act of letting go.

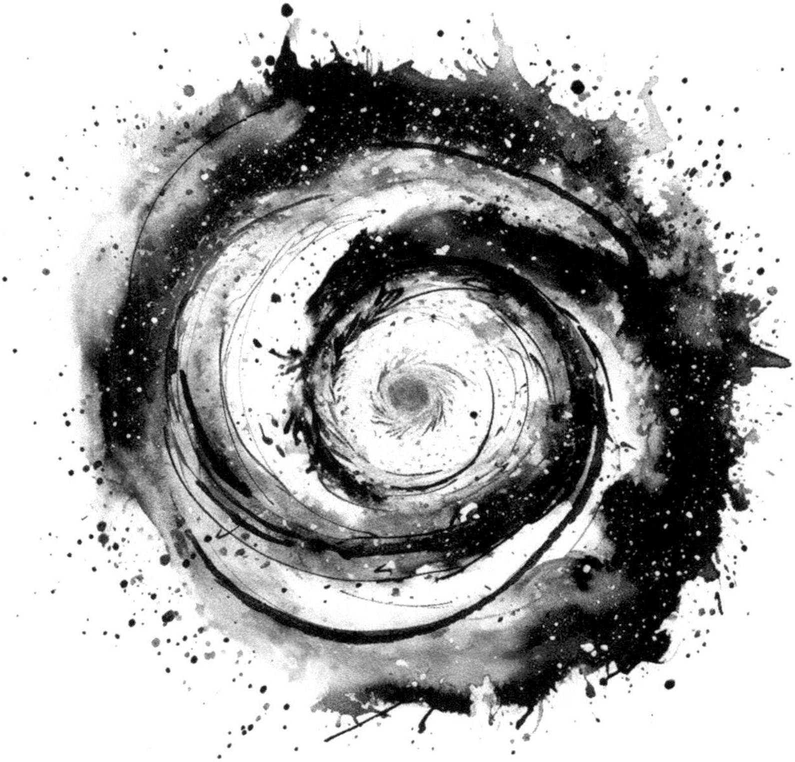

Am I ready now,
for this leap of faith?

Beyond the Brink is the Beginning

Toes curled to grip the edge
This gaping earth
Open cast and ground down
To dust and poison
Head hung with leaden hope
And unrequited grieving

A mile or more into the dusk
A canyon carved
By bloodstained hands
Whose profits soar
On gilded wings
Into the fading sun

Will we take flight, beyond the brink?
Free falling, find our element
And so begin again to see
This untold world we share

Where power lives in every heart
The answers staring eagle eyed,
Each wild thing, its own new start
Our real world unrealised

Reach down and touch this spark of dawning
Palm flat to unyielding earth
The energy we need exists
To rise as one and choose rebirth

We will take flight, beyond the brink
And find a new beginning there
Awakening what lies beneath
The only world we'll ever share

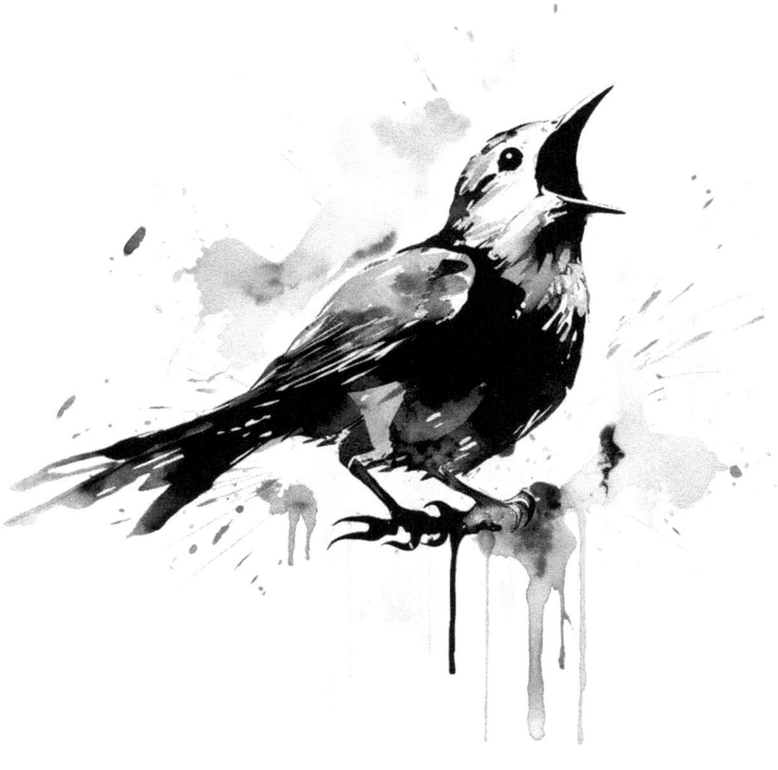

Can you hear
the song of change?

A Song from the Edge of the Universe

Sing from the edge of the universe
Each moment a calling to be
A vision of all that is living
The delta, the lung and the tree

Follow the voice at the centre
The womb and the heart of the earth
Evolving to greet all companions
As guests at our conscious rebirth

Sing softly to dragonfly wing beats
Let the great mother tree be your guide
and dance like the ants and the aphids
to the soul music nature provides

Sing out with the voice that lies dormant
Sing out with your own wild song
Sing out to the roots and the mountains
Let all the lost voices grow strong

Invite oceans to vote for their leaders
Let the forests hold shares in big oil
Let the honeybees plan our cities
Let the earthworms reclaim their soil

Sing at the edge of the universe
At one with the sweet melody
A vision of all that is living
Each moment a calling to be

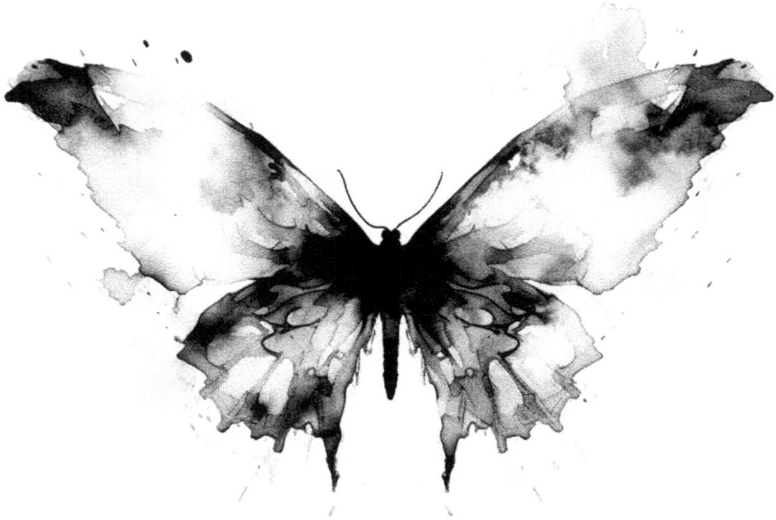

Will I fall or
will you fly with me?

We Share a Dream

We share a dream
You know it too my friend
There's light beyond our common fear
This is not the end

We share a dream
A city full of trees
Ice building on the mountain peaks
Birdsong on the breeze

Farms full of life
And soil rich as gold
The wealth we stand to gain if we
Let nature take a hold

Power in our hands
And constantly renewed
Sun, wind and raging river will
Light this world anew

We share a dream
You know it too my friend
There's light beyond our common fear
This is not the end

There is a fire
Inside you too my friend
The common thread of our desire
A flame that will transcend

This skin and bone
This moment held
This turning point
From which we tell

A tale of peace with every voice
Regenerate with every choice
Unleash the change in every soul
Take up the threads and make this whole
Unravelled world a tapestry
The beauty you were made to be
The beauty of the grass, the tree
The birds in flight, the honey bee

Each ecosystem like each heart
Beat out this rhythm from the start
And so it will again my friend
We share a dream.
It's not the end.

We share a dream
And we can make it true
The answers are all known to us
We know what we must do

This is the time
To find the dream again
Let compassion rise inside us
And with that power exclaim

We are the world
We shall not sit and burn
Each precious conscious being
It is time to take our turn

This skin and bone
This moment held
This turning point
From which we yell

A chant of peace with every voice
Rebuild this world with every choice
Unleash the change in every soul
Take up the threads and make this whole
Miraculous unfolding dream
The purpose of your every scheme

Reality is yours to choose

If all are one we cannot lose

Each human being like each heart
Beat out this rhythm from the start

And so it will again my friend
We have a dream
It's not the end

We will live in a world remade
Your dream and mine will never fade
The best of us will soon be seen

because you and I

We share a dream.

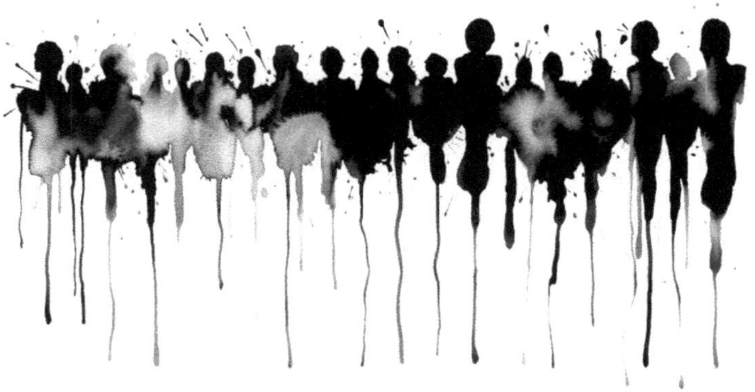

Can I help others
to step forward?

In the Eyes of Another

A life for ourselves
A life for the people we love
A life for all that breathes and sets root in the soil

A life for this fragile planet

Look for this now

In the eyes of another

Smile

Let them see you

And know

That you both stand at the Brink
That there is no turning back
And that real change comes

When you know this in your heart
When you let go of this fractured world

And find your place

Within this web of life

This perfect living Earth

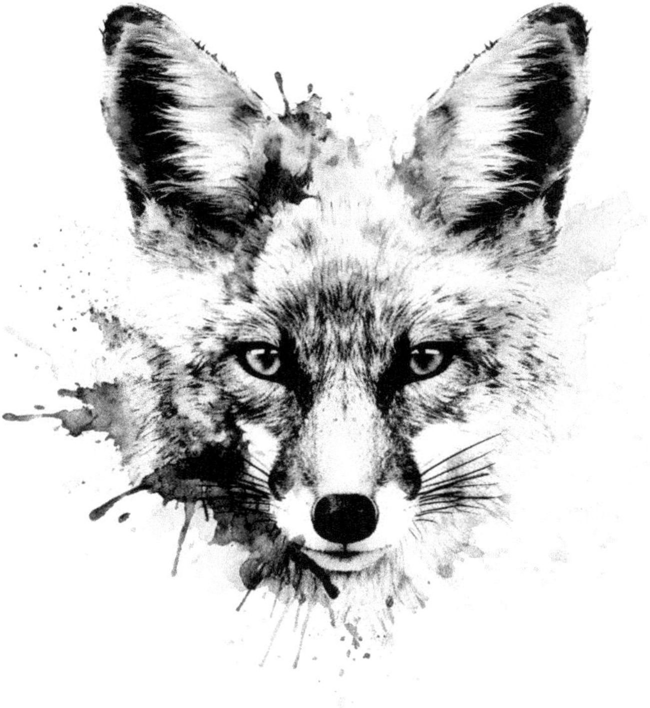

What Next?

None of us can rely on our governments, corporations, bosses, parents or anyone else in a 'position of responsibility' to take care of the climate and biodiversity crises for us. These deeply rooted existential challenges are, in every sense, shared.

The poems in this book are a calling to embrace our collective responsibility towards this planet and the living things around us (human and otherwise). They are an acknowledgement of a beginning and an ending. A world that is being imagined into being and a world that is crumbling before our eyes.

There are many inspirational people working towards our new beginning and they are as responsible as I am for the words in this book. I'd like to tell you a little bit about a few of them because they are beyond the brink already, stepping forward into the light of a new world in which solutions to crises are not just possible, but very real.

I recommend, if you are in search of hope or inspiration, that you find out more about them, read their work, listen to their podcasts and join them in spreading the answers that will bring about the very best of new beginnings.

Travellers Beyond the Brink

This may be a short book, but there are a lot of people who have contributed to its writing and I'd like to introduce you to them.

Manda Scott is a renowned novelist (The Boudica Series and A Treachery of Spies) and podcaster (Accidental Gods). I first came across Manda when I stumbled upon the wonderful Thrutopia writing masterclass, a course which she co-designed to sweep a generation of writers beyond the narrative brink to a plausible, flourishing beginning. I was immediately struck by her passion, intelligence and commitment to writing (and inspiring others to write) stories that lead us into a better future. Many of the people who inspired these poems were gifted to me by Manda and the Thrutopia community. It is fair to say that I would not have written this book without her.

Marcus Link is a poet (Quanticles of Emergence), radical, natural entrepreneur and a dear friend who was once also my business partner. He now runs a regenerative farming business called New Foundation Farms and is also part of a holistic think tank called the Holos Earth Project. Marcus gave a talk for Holos Earth in 2022 on the theme of 'Humans as a beneficial key stone species on earth', which was the inspiration for the poem 'A Song from the Edge of the Universe'.

As part of the Thrutopia community, I was incredibly fortunate to be in the audience for online talks by Rob Hopkins (Co-founder of the Transition network and author of 'From what is to what if') and Jeremy Lent (described by Guardian journalist George Monbiot as "one of the greatest thinkers of our age" and author of 'The Patterning Instinct'). These talks inspired the poem 'The Future is Already Here'.

If you're seeking connection and new ideas, Jeremy has created a very special corner of the Internet called the Deep Transformation Network. The network provides a space to explore pathways to an ecological civilization and you will be welcome there.

Thrutopia talks by B. Lorraine Smith and Roz Savage were the inspiration behind 'Lose Myself' and 'Row Across the Sea'. Roz's book 'The Ocean in a Drop' and B. Lorraine Smith's website (www.blorrainesmith.com) are very much worth a read.

Atoms was inspired by the many talks and writings of the late Thich Nhat Hahn. If you're not familiar with him, I recommend that you read both 'The Art of Living' and 'Zen and the Art of Saving the Planet'. Both of these books have brought me closer to the Brink while helping me to recognise that whatever happens beyond it will happen. I can choose to breathe, to smile, to pay attention.

Suzanne Simard's 'Finding the Mother Tree' took me part way towards writing The Letting Go. The rest took shape during numerous hours spent both wandering and wondering in the woodland that runs along the Yealm Estuary in South Devon.

I've been fortunate enough to encounter a number of other clear voices along this journey. If you are interested in how businesses might adapt beyond the brink, take a look at Paul Skinner's book, The Purpose Upgrade. If you'd like to move away from the consumer story that dominates our lives and find ways to make a positive impact within your community you could start by reading Jon Alexander's book Citizen.

If you can feel the beat of nature in your heart and sense the wholeness of the world but you can't quite reach out and touch it, I'd recommend reading Joanna Macy's Active Hope, the strap line for which (How to Face the Mess We're in Without Going Crazy), summarises my reason for starting this book in the first place.

There is so much to read in this world and so much else to do besides, so I will leave the list at that.

The remaining poems in this book were inspired by events in my life and by quiet time spent in the woods, on the moors and by the sea. For these, I'd like to thank the more than human world, the living, the breathing, the ever evolving, from which we borrow all things.

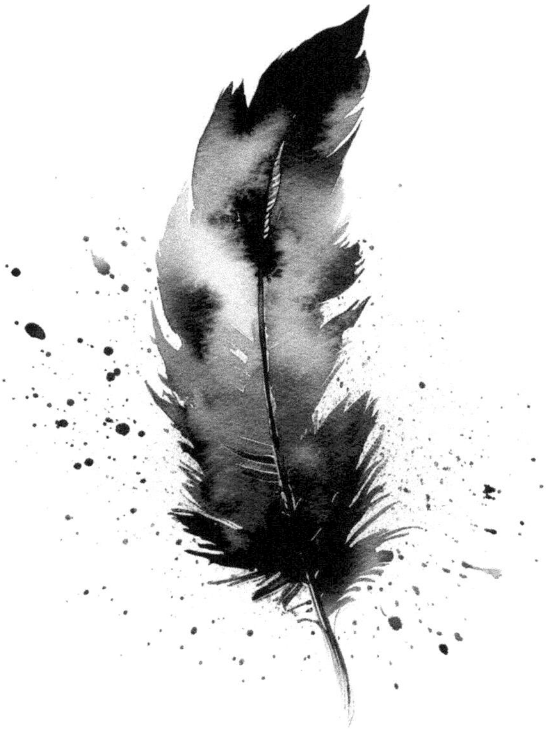

Acknowledgements

A few other people deserve a mention for their support, encouragement and wisdom. In no particular order, Calvin Niles, David Buckley, Dom Cooper, Helen Bennett, Richard Hagen, Mary Ann Allison, Ashley Woodhall, Amanda Keatley, Vanessa Cobb, Bob Greig, Rebecca Giraud and all the members of the Thrutopia Writing Group.

I greatly appreciate the time spent by my parents, who have suffered my poetry longer than most and who are always happy to read, edit and say nice things about me.

Finally, I am constantly and joyfully grateful for the smiles and laughter of my children and the unbreakable support of my wife (of twenty years already) Susanna. My family has provided more than enough fuel for this project and many others.

About the author

I wonder, regularly, if it isn't entirely misguided for someone like me (That guy, white, British, male, middle-class, middle-aged, straight, no notable achievements, no celebrity, never been on Strictly, never even watched Strictly actually!) to write this stuff at all, especially a collection of poetry! I mean what sort of person wakes up at 43 and thinks, what the world needs now is more rhyming couplets from the British middle-classes?

Well, this sort, apparently!

So far, mine has been a meandering path. There have been many moments of luck and not very much of a plan. I try to recognise the privilege of living this way and to be thankful for it. I've had a go at being a rock star, a computer scientist and a primary school teacher. I've run a few small businesses, packaged pottery, cleaned the kitchen at a care home, the list goes on. I've been fortunate to pick up a few different perspectives along the way and to be guided by many clearer voices than my own.

I grew up on the edge of Dartmoor, surrounded by nature and spent the first nineteen years of my life within a stones throw of the beautiful South Hams coast and the rugged hills. This was, again, a great privilege that went largely unrecognised until I took a ten year break from Devon and realised just how much I missed it.

I now live and work in a small village close to the coast with my wife and two sons. I can walk to the woods and to the estuary. If I'm lucky I might see a Woodpecker, a Dipper or on the rarest of rare occasions an Osprey.

The wind whistles through the trees,
the tide rises and falls.

I am, occasionally, uncertain, freaked out and deeply unsettled about the future. However, there are times when I am at peace, content in the knowledge that this tiny world, bathed in a blanket of a trillion stars, is astonishing, and that we are, after all, just a tiny part of it.

This is not
The End

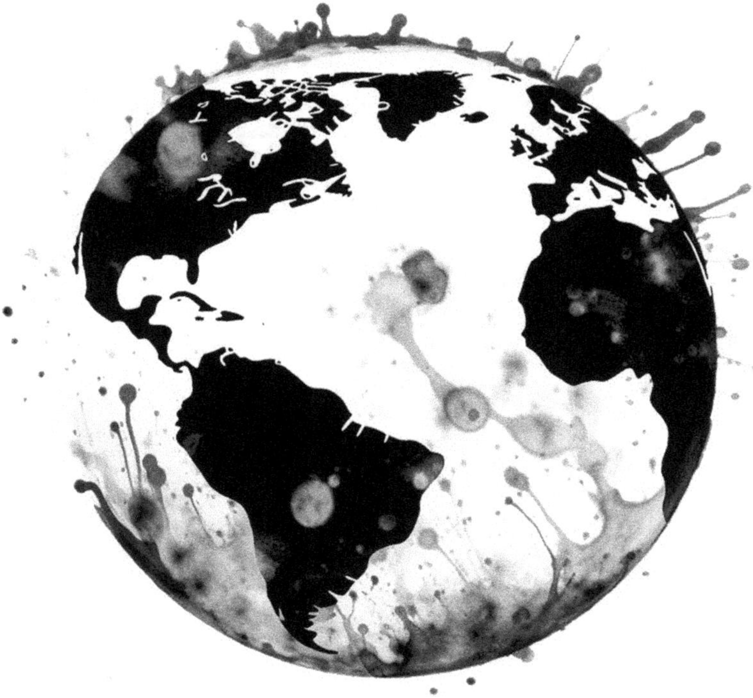

Looking for more poems?

You can listen to more of Richard's poems by subscribing to his YouTube Channel or following him on Instagram @richardwainpoet

Supporting Devon Environment Foundation

£1 from every sale of this book will be donated to the Devon Environment Foundation to support their work protecting and restoring Devon's natural beauty by funding local nature-based solutions.